ULTIMATE SUPERCARS

PORSCHE 911 TURBO S

By Joanne Mattern

Kaleidoscope
Minneapolis, MN

The Quest for Discovery Never Ends

This edition first published in 2023 by Kaleidoscope Publishing, Inc.

No part of this publication may be reproduced in whole or in part without written permission of the publisher.

For information regarding permission, write to
Kaleidoscope Publishing, Inc.
6012 Blue Circle Drive
Minnetonka, MN 55343

Library of Congress Control Number
2022938005

ISBN
978-1-64519-613-6 (library bound)
978-1-64519-683-9 (ebook)

Text copyright © 2023 by Kaleidoscope Publishing, Inc. All-Star Sports, Bigfoot Books, and associated logos are trademarks and/or registered trademarks of Kaleidoscope Publishing, Inc.

Printed in the United States of America.

FIND ME IF YOU CAN!

Bigfoot lurks within one of the images in this book. It's up to you to find him!

TABLE OF CONTENTS

Chapter 1: First Look 4

Chapter 2: Looking Back 12

Chapter 3: Starting and Stopping 18

Chapter 4: Fastest on the Track 24

Beyond the Book.. 28
Research Ninja... 29
Further Resources... 30
Glossary... 31
Index... 32
Photo Credits... 32
About the Author... 32

Chapter 1
First Look

Rose was waiting by the front door when her father came home from his business trip. "Hello, Dad," she said, giving him a big hug. "How was the car show?"

"Great!" her father said. "I brought you a present." He reached into his bag and pulled out a small toy car. "This is the newest Porsche 911 Turbo S. Sorry I couldn't buy you the real thing."

Rose laughed. Her father handed her a booklet. "Here is more information about the car. I think you'll enjoy reading this."

"Thanks!" Rose said. She sat down on the couch and began to look through the book. She turned to the pages about the engine. "Ooh, the car has a **flat-six** engine with a twin turbo drive," she read. "I'll bet that creates a lot of power and speed."

"You bet," her father said as he sat down beside her. "The engine produces 640 **horsepower**. And the **turbochargers** are bigger than they were in earlier models. That gives the driver more power and control."

"Listen to this," Rose said. "The 911 can have a sport **exhaust** system. This gives the car a more powerful sound."

FUN FACT
The "S" in the car's name stands for "Super" or "Sport." An S model has a more powerful engine than other models.

"That's great," her father said. "There is nothing like the sound of a powerful racing engine."

Rose kept reading. "This says Porsche put a new **transmission** into the 911 Turbo S. It lets the driver change gears more quickly." She sighed. "I wish I could drive this car. It sounds and looks amazing!"

"Well, it certainly is fast," her father said. "Top speed on the track is 205 miles per hour (330 km/h).

"How much does the car cost?" Rose asked.

"Only $207,000," her father said. "That's a bit too much for our budget. But it is fun to dream, isn't it?"

Rose stared at the beautiful car in the book. "It sure is," she said.

AIR FLOW

Car engines need air to burn fuel and make the car work. Getting the right amount of air into the engine is very important. The 911 used to have air intakes on the sides of the car. The 2022 model has air intakes in the back. This allows more air to flow into the engine and gives the engine more power.

PARTS OF A
911 TURBO S

rear wing

light strip

Sport Cup tires

Chapter 2
Looking Back

Porsche's 911 model actually started as the 901. In 1963, Ferry Porsche introduced a new kind of car. Like other sports cars, this car had an air-cooled engine in the back. But Porsche gave the car a more powerful six-cylinder engine called a boxer.

Porsche showed the 901 at an auto show in Frankfurt, Germany. The car was a big hit. However, the French car company Peugeot complained that the car's name was the same as their own 901 model. So Porsche changed the 901 to the 911.

FUN FACT
Ferry Porsche's son, Ferdinand III, designed the body of the 901.

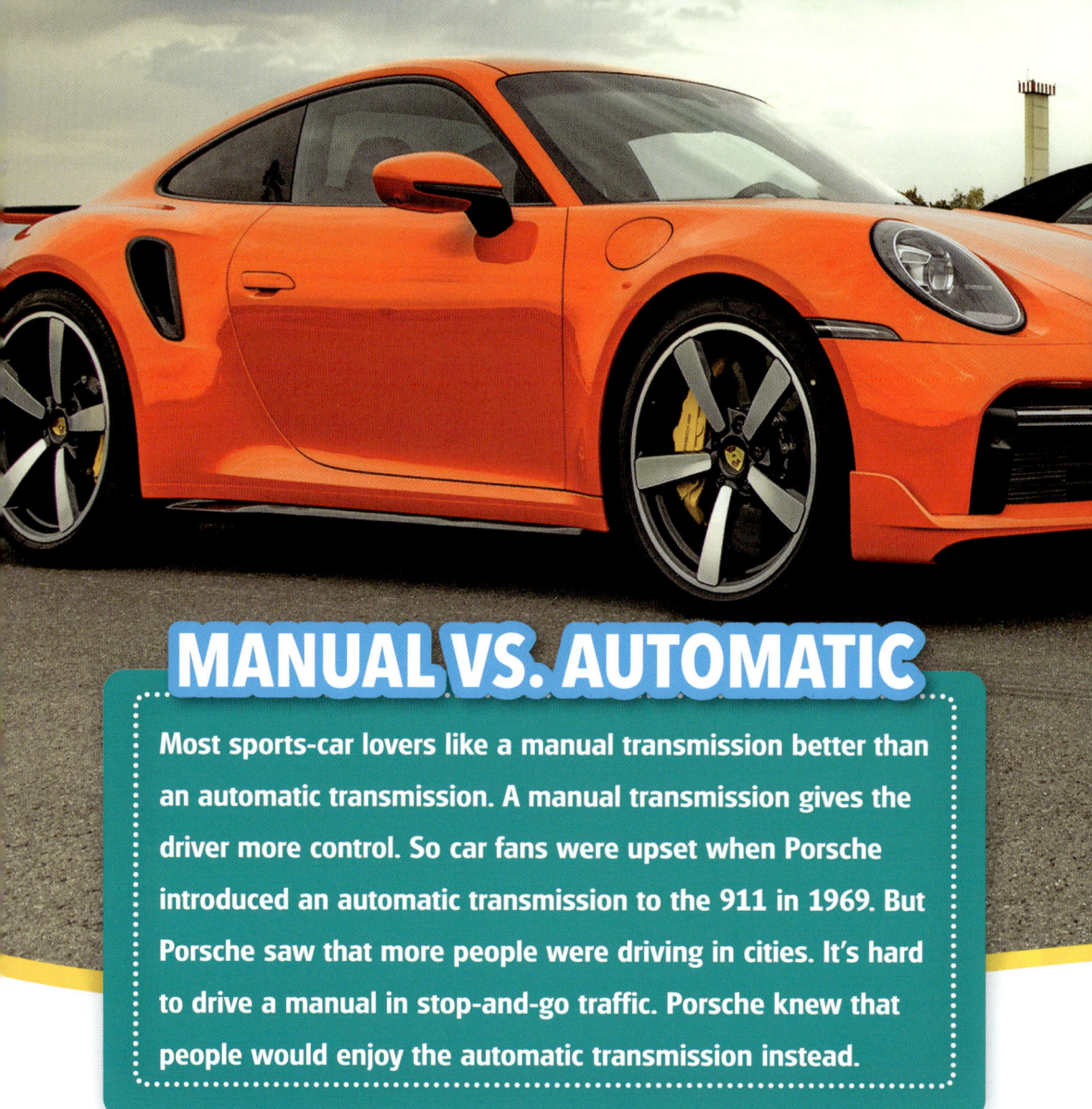

MANUAL VS. AUTOMATIC

Most sports-car lovers like a manual transmission better than an automatic transmission. A manual transmission gives the driver more control. So car fans were upset when Porsche introduced an automatic transmission to the 911 in 1969. But Porsche saw that more people were driving in cities. It's hard to drive a manual in stop-and-go traffic. Porsche knew that people would enjoy the automatic transmission instead.

Porsche made new models of the 911 every few years. Each model gave the car more power. Engines got bigger. So did the car's horsepower. Porsche used materials like aluminum bumpers and wheels to make the car lighter. This also increased speed.

FUN FACT
The first 911s were sold in the United States in 1965.

Porsche introduced the first 911 Turbo in 1975. The Turbo model had a special pressure control in the exhaust system. This control helped the car's engine work better and gave it more power. At that time, the 911 Turbo was the fastest car in Germany.

In 1992, the first 911 Turbo S appeared. This car was sportier than other models. The car was low to the ground and very fast. But Porsche only made 80 of these cars.

By the early 2000s, the 911 looked very different than earlier models. The car was long and sleek instead of boxy. This design looked great. It also made the car more **aerodynamic**. That meant the car could go faster!

Porsche always wanted to create a car that could perform on the racetrack but also be fun to drive on streets and highways. The Porsche 911 Turbo S does just that.

WHERE IS THE PORSCHE 911 TURBO S MADE?

GERMAN MADE

The Porsche 911 Turbo S is made at Porsche's factory in Zuffenhausen, Germany. Porsche's headquarters are also in Zuffenhausen. Their research and development center is in Weissach, Germany.

Chapter 3
Starting and Stopping

The 2022 Porsche 911 Turbo S has some amazing features. One of the coolest is the car's Sport Modes. A Sport Response button on the steering wheel gives drivers a choice of five different driving modes. These modes are Normal, Sport, Sport Plus, Individual, and Wet.

Normal mode is perfect for everyday driving. Sport mode lets the car respond more quickly to steering and acceleration.

FUN FACT
The front spoiler and rear wing on the 911 Turbo S can move out. This helps slow down the car when braking at high speeds.

The car handles even better in Sport Plus mode. Rear-axle steering makes the car easier to control and more stable going around curves and corners. Individual mode lets drivers pick their favorite features. They can create a ride that is exactly what they want or need.

Those aren't all the features found in the 911 Turbo S. The car's Sport Chrono Package has three more tricks up its sleeve. The first is Launch Control. This feature gets the car off to the fastest start.

The second feature changes the way the gears shift. The shift time is shorter. This allows the car to **accelerate** faster.

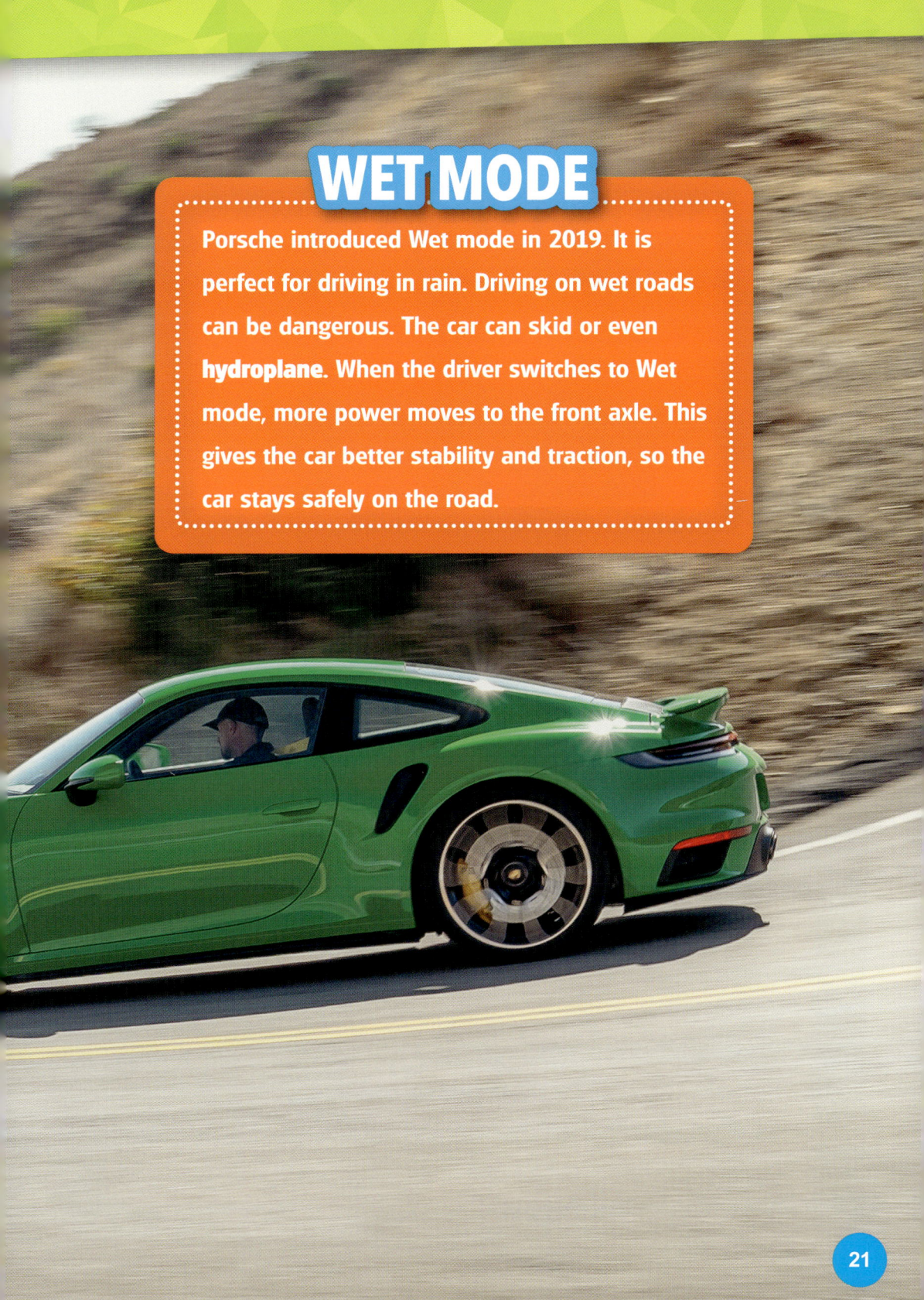

WET MODE

Porsche introduced Wet mode in 2019. It is perfect for driving in rain. Driving on wet roads can be dangerous. The car can skid or even **hydroplane**. When the driver switches to Wet mode, more power moves to the front axle. This gives the car better stability and traction, so the car stays safely on the road.

THE 911 TURBO S
IN DETAIL

Height: 4.3 feet (1.3 m)

Width: 6.2 feet (1.9 m)

LENGTH: 14.9 feet (4.5 m)

WEIGHT: 3,636 pounds (1,649 kg)

TOP SPEED: 205 mph (330 kph)

TIME FROM 0 to 60 miles per hour: 2.6 seconds

This third feature allows the engine and transmission to produce a huge amount of power. This burst of power only lasts about 20 seconds. That's enough to send the car flying down the road or track.

Of course, even the fastest car needs to stop quickly and safely. The 911 Turbo S features Porsche Ceramic Composite Brakes, or PCCBs. These brakes are lighter than cast-iron brakes. They are also a little larger. These features help the car stop more smoothly. They also help the car hold the road better.

Chapter 4
Fastest on the Track

Nick walked around the Porsche 911 Turbo S. He liked what he saw. The car was sleek. It looked fast and powerful. He had no doubt it would win the race.

A car magazine had set up some drag races between different cars. Nick would test the Turbo S. He would race against a few other sports cars. The first was an Audi R8 V-10.

FUN FACT

The 911 Turbo S has a new 8-speed manual transmission called PDK. It allows drivers to change gears very quickly without any hesitation in power.

It was time for the first run. Nick shook hands with the other driver. Then he climbed into the 911 Turbo S and gripped the steering wheel.

The car started with a roar. Nick liked the throb of the engine. The light turned green. Nick shifted into first and pushed the Launch Control button. The Porsche leaped forward.

The 911 zoomed down the track. Nick loved the feel of the car. He was in total control, despite the high speed.

The quarter-mile race went by in a flash. Nick was ahead of the Audi the whole time. When the checkered flag came down, Nick climbed out of the car. The Audi pulled up next to him.

"How did I do?" Nick asked.

"10.1 seconds for you. That's much faster than the 11 seconds the Audi did."

Nick laughed. That was good news. But the day got even better.

Next, the Porsche matched up with a Nissan GT-R Nismo. Once again, Nick had the upper hand. He roared across the finish line in 10.1 seconds. The Nissan took 10.8 to race the quarter mile.

"It's amazing," Nick said. "I could drive this car home on the streets with no problem. Yet it is a beast on the racetrack. The Porsche 911 Turbo S really is the best of both worlds."

A TIGHT FIT

Most supercars are not very big. Being small makes these cars faster! The Porsche 911 Turbo S can hold four people. However, the two seats in the back are so small, only children can fit back there. Adult passengers are out of luck.

BEYOND THE BOOK

After reading the book, it's time to think about what you learned. Try the following exercises to jump-start your ideas.

THINK

DIFFERENT SOURCES. Think about types of sources you could find on the Porsche 911 Turbo S. What could you find in a magazine? What could you learn at a dealership? How could each of the sources be useful in its own way?

CREATE

GET ARTISTIC. Cars start with creative artists and designers. Time for you to take a shot! Get art materials and create a great, new car. Will you make it a sports car? A sedan? A race car? What colors will you paint it? What features can you give it? Let your imagination go for a spin!

DISCOVER

SHARPEN YOUR RESEARCH SKILLS. The Porsche 911 Turbo S is turbocharged. Where could you go in the library to find more information about turbocharged motors in cars? Who could you talk to who might know more? Create a research plan. Write a paragraph about your next steps.

GROW

GO TO A CAR SHOW. Car shows are a great way to see lots of cool cars up-close. Check your local events calendar, or ask at a car dealer for upcoming events. You can find shows of old cars and new cars, sports cars and classic cars. Go to a show and find a new favorite car to love!

RESEARCH NINJA

Visit **www.ninjaresearcher.com/6136** to learn how to take your research skills and book report writing to the next level!

RESEARCH

DIGITAL LITERACY TOOLS

SEARCH LIKE A PRO
Learn about how to use search engines to find useful websites.

FACT OR FAKE?
Discover how you can tell a trusted website from an untrustworthy resource.

TEXT DETECTIVE
Explore how to zero in on the information you need most.

SHOW YOUR WORK
Research responsibly— learn how to cite sources.

WRITE

GET TO THE POINT
Learn how to express your main ideas.

PLAN OF ATTACK
Learn prewriting exercises and create an outline.

DOWNLOADABLE REPORT FORMS

Further Resources

BOOKS

Cockerham, Paul. *Porsche: The Ultimate Speed Machine*. Broomall, PA: Mason Crest, 2018.

Murray, Julie. *Porsche 911*. Minneapolis, MN: Abdo Zoom, 2018.

Mason, Paul. *German Supercars: Porsche, Audi, Mercedes*. New York: PowerKids Press, 2019.

WEBSITES

FACTSURFER

Factsurfer.com gives you a safe, fun way to find more information.

1. Go to www.factsurfer.com.
2. Enter "Porsche 911 Turbo S" into the search box and click 🔍
3. Select your book cover to see a list of related websites.

Glossary

accelerate: to accelerate means to go faster. The Porsche 911 Turbo S can accelerate from 0 to 60 in just 2.6 seconds.

aerodynamic: an aerodynamic design reduces the drag, or pull, on a car as it moves through the air. The Porsche 911 Turbo S's shape makes it more aerodynamic.

exhaust: a car's exhaust has pipes that release waste gases into the car. Some cars have very loud exhausts.

flat-six: a flat-six engine has 6 cylinders. Three cylinders are on each side of the crankshaft. The Porsche 911 Turbo S has a flat-six engine.

horsepower: horsepower measures the power of the engine. The Porsche 911 Turbo S has 640 horsepower.

hydroplane: if a car hydroplanes, it slides on water covering the road. It is very hard to control a car when it hydroplanes.

transmission: the transmission is the part of the car that moves power from the engine to the wheels. The Porsche 911 Turbo S has an eight-speed automatic transmission.

turbocharger: a turbocharger is a supercharger powered by gases from the car's exhaust. A turbocharger can make a car go much faster.

Index

air intakes, 8
engine, 4, 6, 7, 8, 12, 14, 15, 23, 25,
exhaust, 6, 15
gears, 7, 20, 24
horsepower, 6, 14
Sport Modes, 18
transmission, 7, 14, 23, 24
turbochargers, 6

PHOTO CREDITS

The images in this book are reproduced through the courtesy of: Porsche Media (Right Light Media 3, 6-7, 18-19; Rossen Gargolov 16; Mark Fagelson Photography 22).
Cover: Porsche Media, slhy/Shutterstock (background).

About the Author

Joanne Mattern has written many nonfiction books for children. Her favorite topics include sports, biographies, animals, and history. Joanne lives in New York State with her family and loves to drive fast cars.